# WHAT IS THIS THING CALLED LEAN?

## Stories of Continuous Improvement

Javier Olivares

## Dedication:

To my beloved wife, Yissel Puello, and our little brave explorer, Santiago Olivares. This work is dedicated to you, who fill my life with love, inspiration, and joy. Thank you for your unwavering support and for being my constant source of motivation.

## Prologue

Continuous improvement is a fascinating journey, one that takes us from concern about the current state to the constant pursuit of excellence. In this book, "What is this thing called Lean? Stories of continuous improvement", we will immerse ourselves in the exciting world of Lean, a methodology that has transformed countless organizations worldwide and has left a deep mark on my own life. Throughout these pages, I will share with you eight inspiring stories that will take us through different industrial sectors, from poultry farming to autoparts manufacturing, and show us how the Lean philosophy can significantly influence how we operate, improve, and thrive in our endeavors.

Each story will present us with a unique set of challenges and obstacles that courageous individuals and committed teams faced. We will see how, through the application of Lean principles, they were able to overcome hurdles, improve efficiency, raise quality, and achieve levels of excellence that seemed unattainable at first. But this book is more than just a collection of stories; it is an invitation to reflection and action. As we navigate through these pages, I encourage you to think about how Lean principles could be applied in your own lives and organizations. Together, we will learn that continuous improvement is not just about tools and techniques but about a mindset, a culture, and a commitment to always seek a better way.

I am grateful to my beloved family, my wife Yissel, and my son Santiago, for their unwavering support on this journey. Also, to my parents, siblings, and all those who have influenced my life and inspired me to share these stories with the world. May this book inspire and motivate you to embrace continuous improvement in all areas of your lives. As the old Lean proverb says, "There is no end to improvement." May these stories remind you that, no matter how challenging the obstacles, there is always a way to move towards excellence.

With enthusiasm and gratitude,

Javier Olivares

**Content**

## Introduction

In a constantly evolving business world, the pursuit of excellence and efficiency has become an obsession. Organizations face the pressure to deliver high-quality products and services, reduce costs, and meet the growing expectations of customers. During this challenge, a philosophy has emerged that has transformed how companies operate, improve, and thrive: Lean.

Lean, a methodology that originated in the automotive industry and later expanded to various sectors, has proven to be a revolutionary force in the quest for continuous improvement. Its focus on eliminating waste, standardizing processes, and prioritizing customer value has led to a paradigm shift in business management. This book, "What is this thing called Lean?", will immerse you in the exciting world of Lean methodology through a series of inspiring stories. Throughout these pages, we will explore how individuals and teams faced unique challenges in industries as diverse as poultry farming, autoparts manufacturing, pastry and customer service.

Through these stories, you will discover how Lean principles were successfully applied to improve operational efficiency, enhance the quality of products and services, and create a culture of continuous improvement. Each narrative will take you on a journey of business transformation and show you how Lean concepts can be adapted and applied in different contexts. But this book is more than just a collection of business stories; it is a guide that will provide you with practical tools and knowledge to implement Lean in your own organization or career. As we progress on this journey, we invite you to explore the fundamentals of Lean, from waste elimination to the creation of efficient value streams.

Prepare yourself for a journey of inspiration and learning that will change your perspective on continuous improvement and equip you with the skills needed to drive success in any industry. Lean is more than a methodology; it is a philosophy that will help you unlock the hidden potential within your organization and within yourself.

So, are you ready to discover what this thing called Lean is all about? Let's embark on this exciting journey together!

### Story 1: "Optimizing Production on a Poultry Farm"

In a small family-owned poultry farm nestled in the picturesque hills of the region, Juan Martínez faced a challenge that kept him awake at night. As the owner of "Happy Birds Farm," Juan had a deep passion for raising high-quality chickens. However, his business was grappling with a series of issues that threatened its viability.

Juan had inherited the farm from his grandfather, who had taught him the secrets of raising healthy and contented chickens. For years, the farm had been known for its fresh and flavorful chicken meat, and many local customers had become loyal buyers. But in recent years, the farm's efficiency had declined, and production costs had skyrocketed.

The most pressing challenge Juan faced was inefficiency in the chicken feeding process. The traditional method they had used for generations involved workers walking through the pens with large bags of feed, manually distributing the food. This was not only labor-intensive but also resulted in significant food wastage, as not all chickens ate at the same time.

Juan knew that something had to change if he wanted to keep the farm alive and preserve his grandfather's legacy. That's when he came across the Lean methodology in a book, he read during a business trip. The idea of eliminating waste and optimizing processes resonated with him. He decided it was time to apply those principles to his farm.

**The Challenge:** Reducing Waste and Increasing Efficiency in Chicken Feeding

The first step Juan took was to gather his team and share his vision for improving the farm. He explained the Lean methodology and how it could be applied to their business. All team members were willing to try something new and accepted the challenge.

The next crucial step in implementing Lean on the farm was to conduct an in-depth analysis of the chicken feeding process. They observed every step, measured times, and identified areas of waste. What they found was revealing: the time spent walking through the pens with bags of feed and the lack of control over the amount of food dispensed were the primary issues.

**The Lean Solution:** Implementing a Continuous Flow System

Based on Lean principles, Juan and his team developed a new chicken feeding system. Instead of walking through the pens with bags of feed, they installed a series of pipes and automatic feeders that dispensed food consistently and controlled. This eliminated the need for workers to walk back and forth and reduced food wastage.

Furthermore, they implemented a real-time monitoring system that allowed them to track the food consumption in each pen and adjust the amount precisely. This ensured that all chickens received the right amount of food, improving uniformity in growth and reducing waste.

**Amazing Results:** Increased Production, Reduced Waste, and Healthier Chickens

The results of implementing Lean on the farm were astonishing. Chicken production increased significantly due to more uniform growth and reduced food waste. Feeding costs decreased by 20% in the first year, contributing to a substantial improvement in profitability.

But it wasn't just about the numbers. The chickens on the farm were healthier than ever. Uniform growth translated into more consistent meat quality, making customers even more satisfied. Additionally, the farm team was more motivated than ever, knowing they were doing a more efficient and sustainable job.

**Lessons Learned:** The Importance of Continuous Improvement

Juan's story and his farm are an inspiring example of how the Lean methodology can be successfully applied in any environment, even on a poultry farm. Juan learned that continuous improvement is essential to maintaining the viability of his business and preserving his family's legacy.

In just a few years, the farm went from struggling to survive to becoming a model of efficiency and quality in chicken farming. Implementing Lean not only improved the farm's profitability but also strengthened customer relationships and increased team satisfaction.

The story of "Happy Birds Farm" serves as a reminder that continuous improvement is an ongoing process. Juan and his team continue to search for ways to optimize their operations and stay at the forefront of best practices in chicken farming. Their commitment to excellence in food production is an example for other farmers and entrepreneurs, demonstrating that even in the most traditional settings, Lean can make a difference.

**Conclusion:** A Bright Future for "Happy Birds Farm"

Today, "Happy Birds Farm" has become a benchmark in raising high-quality chickens. Juan Martínez has shared his success story in continuous improvement conferences and workshops, inspiring other farmers to apply Lean principles in their operations.

The farm has not only survived but thrived and is looking toward a bright future. Juan's story and his poultry farm are a reminder that, regardless of the industry you are in, continuous improvement can lead to astonishing results. With dedication, vision, and a steadfast commitment to excellence, any business can transform and prosper in an increasingly competitive world.

**Story 2: "Waste Reduction at a Poultry Processing Plant"**

On the outskirts of a small town stood the poultry processing plant, "AviProcessa," a company dedicated to producing high-quality chicken and turkey meat. For years, they had been a pillar in the community, providing employment and food products to the region. However, the company was grappling with a problem that had been growing over time: the amount of waste generated during the poultry processing process.

Maria Rodriguez, the operations manager at AviProcessa, had worked at the plant for over a decade. She had seen the company grow and expand, but she had also seen issues related to waste management becoming increasingly prominent. Piles of feathers, entrails, and other byproducts of production were accumulating rapidly, leading to significant disposal costs and environmental concerns.

**The Challenge:** Increasing Waste and Disposal Costs

The most pressing challenge Maria faced was the amount of waste generated at the plant. As production grew to meet market demand, the accumulation of feathers and entrails also grew, resulting in a greater need for storage space and disposal costs.

In addition to the economic issues, Maria was also aware of the company's environmental responsibility. AviProcessa had been operating in the same location for decades, and the surrounding community relied on the health of its natural environment. The buildup of waste was not only a financial problem but also a threat to the company's reputation and its relationship with the local community.

## The Lean Solution: Identifying and Eliminating Waste

Faced with these challenges, Maria decided to act. After researching various continuous improvement methodologies, she chose to apply Lean principles at the plant. The philosophy of eliminating waste and optimizing processes resonated with AviProcessa's situation.

The first step was to form a continuous improvement team composed of employees from different areas of the plant. Maria believed in the importance of involving all levels of the organization to succeed in any improvement project.

Next, the improvement team conducted a detailed analysis of the production process and waste management. They observed every step of the process, from receiving the birds to packaging the final products. They recorded how feathers and entrails were handled, stored, and transported for disposal. What they discovered was that there were multiple opportunities to reduce waste. They realized that a significant number of feathers and entrails could be reused or recycled in some way rather than being completely disposed of. One of the key changes they implemented was a waste segregation system. Instead of mixing all the waste in one container, separate storage areas were established for different types of waste, such as feathers, entrails, and other byproducts. This allowed for more efficient and effective waste management.

Another significant initiative was to seek opportunities for recycling waste. They partnered with local companies that could turn feathers into useful products, such as pillows and quilt fillings. The entrails were directed towards the production of pet food and organic fertilizers. These initiatives not only reduced disposal costs but also generated additional income for the company. AviProcessa began selling products derived from waste, contributing to the improvement of its financial situation.

**Amazing Results:** Cost Reduction and Enhanced Reputation

The results of implementing Lean at AviProcessa were remarkable. In the first year, the plant managed to reduce waste disposal costs by 40%. This translated into a significant increase in the company's profitability.

In addition to the financial benefits, AviProcessa's reputation improved significantly in the local community. The plant became a model of environmental responsibility and sustainability. Nearby residents no longer saw heaps of waste piling up on-site, and the company began receiving praise for its commitment to the environment.

**Lessons Learned:** The Importance of Sustainability

The story of AviProcessa is a testament to the power of continuous improvement and sustainability in business. Maria and her team demonstrated that even in a traditional industry like poultry processing, it is possible to reduce waste and improve profitability simultaneously.

Efficient waste management not only helped AviProcessa reduce costs and generate additional income but also strengthened its position in the community and its reputation in the market. The company went from being an environmental problem to becoming a model of responsibility.

**Conclusion:** A Sustainable Future for AviProcessa

Today, AviProcessa takes pride in its commitment to sustainability and continuous improvement. The company has set higher standards in the industry in terms of waste management and environmental responsibility.

Maria's story and AviProcessa demonstrate that sustainability is not just an ethical goal but also a smart business strategy. Waste reduction and efficient resource management can generate economic benefits and improve a company's reputation.

AviProcessa continues to look for ways to improve its process and further reduce waste, showing that continuous improvement is an ongoing journey toward a more sustainable and profitable future. The story of the poultry processing plant is a reminder that, regardless of the industry, applying Lean principles and sustainability can lead to remarkable results. With dedication and a long-term vision, any company can transform and thrive in an increasingly environmentally conscious world.

## Story 3: "Reducing Waste at an Artisanal Bakery"

In the peaceful village of San Lorenzo, nestled in the heart of the poultry region, stood the family farm "Success Birds." For decades, this farm had been known for producing high-quality poultry. However, as time marched on, challenges arose that threatened the sustainability of the business. It was amidst these difficulties that Roberto Gomez, the farm's owner, decided to embark on a journey of transformation.

The poultry farm had been founded by Roberto's grandparents and had been passed down from generation to generation. Despite the rich tradition and pride surrounding the business, Roberto faced a challenge that jeopardized its survival: production efficiency. As the demand for chicken and other poultry products increased, the farm found itself at a crossroads.

**The Challenge:** Production Efficiency and Product Quality

The primary challenge Roberto faced was production efficiency. The farm operated using traditional methods that had been handed down through generations, but these methods were no longer sufficient to meet market demand and keep costs low. Additionally, product quality was gradually declining due to the lack of control and standardization in production processes.

Competition in the poultry market was fierce, and Roberto realized that if he didn't improve the efficiency of his farm, he risked losing customers and eventually the business he loved so much. That's when he decided it was time to explore new ways of doing things.

**The Lean Solution:** Applying Continuous Improvement Principles

Roberto had heard about the Lean methodology through a friend who had used these principles to improve his logistics company. After researching Lean further and how it could be applied to poultry production, Roberto was convinced that this philosophy of continuous improvement could be the key to addressing his farm's challenges.

The first step he took was to gather his team and explain his vision for improving the farm. He introduced them to Lean principles and how they could be applied to poultry production. While some team members were initially skeptical, they were willing to try something new for the sake of the farm.

The next step was to conduct a detailed analysis of the production process on the farm. The team observed every step, from the arrival of the chicks to the packaging of the final products. They meticulously recorded production times, identified bottlenecks, and looked for areas where unnecessary time and resources were being wasted.

What they discovered was enlightening. There were many opportunities to improve efficiency on the farm. For instance, they noticed that the chicken feeding process was disorganized, and a significant amount of feed was being wasted. Additionally, the cleaning and disinfection times of the facilities were not optimized, causing delays in production.

With data and observations in hand, the Aves de Exito improvement team began implementing significant changes on the farm. One of the initial changes was the reorganization of the chicken feeding process. Instead of feeding the chickens traditionally, an automated feeding system was implemented, which accurately dispensed food and eliminated waste.

Furthermore, a visual management system was established throughout the farm to help workers identify issues and react quickly. This included key performance indicators that displayed real-time progress and allowed for informed decision-making.

The team also focused on staff training. Training programs were created to enhance the skills and knowledge of workers in critical areas, such as poultry handling and facility cleaning. Training not only improved the quality of work but also boosted team morale.

**Amazing Results:** Increased Efficiency and Product Quality

The results of implementing Lean on the farm were astounding. Poultry production increased significantly due to improved feeding efficiency and reduced food waste. Moreover, product quality significantly improved thanks to process standardization.

In the first year of Lean implementation, Aves de Exito managed to reduce production costs by 15%. This not only improved the farm's profitability but also allowed them to maintain competitive prices in the market.

**Lessons Learned:** The Importance of Innovation and Continuous Improvement

Roberto and Aves de Exito story is a testament to the power of innovation and continuous improvement in any industry, even in traditional poultry production. Roberto learned that, to remain competitive and sustainable, it was necessary to challenge entrenched practices and constantly seek ways to improve.

Over the years, Aves de Exito has continued to apply Lean principles and seek ways to innovate its production process. Roberto and his team have shown that continuous improvement is an ongoing journey, and they are committed to preserving the farm's legacy for many more generations.

**Conclusion:** A Bright Future for Aves de Exito

Today, Aves de Exito is a model of efficiency and quality in poultry production. The farm has maintained its position as a leader in the region and has expanded its presence in the national market. Roberto's story and his poultry farm serve as a reminder that innovation and continuous improvement can lead to remarkable results, even in traditional industries. With vision, commitment, and a steadfast focus on excellence, any business can transform and thrive in an increasingly competitive world.

**Story 4: "Improving Efficiency at a Boutique Bakery"**

In the heart of the city, on a picturesque corner, stood the boutique bakery "Sweet Dreams." This small bakery was known for its exquisite cakes, delicious desserts, and a cozy atmosphere that attracted customers from all around. However, behind the charming façade and tempting aromas, the bakery faced challenges that threatened its ability to thrive in a competitive market.

Laura Torres, the owner and head pastry chef of Sweet Dreams, had a passion for baking that she had inherited from her grandmother. For years, she had run the bakery with love and dedication, creating delights that delighted her customers. However, as the bakery's popularity grew, so did the operational challenges.

**The Challenge:** Production Efficiency and Delivery Times

The main challenge Laura faced was production efficiency and the management of delivery times. Sweet Dreams had earned a reputation for its custom cakes, but an increasing number of customers complained that orders were not delivered on time. Laura struggled to maintain the quality of her products, but the growing demand and disorganized processes were affecting her ability to meet delivery deadlines.

The bakery also faced cost issues. High-quality ingredients and trained staff were expensive, and Laura was struggling to maintain profit margins while meeting her customers' expectations.

**The Lean Solution:** Focusing on Efficiency and Organization

Laura had heard about the Lean methodology through a friend who had applied these principles in his manufacturing business. Although Lean initially seemed to focus on manufacturing goods, Laura was convinced that Lean principles could also be applied to the bakery to address its operational challenges.

She decided it was time to act and apply Lean principles to Sweet Dreams. The first step was to gather her team and share her vision of improving efficiency and quality in the bakery. All team members were willing to try something new to help the bakery grow and thrive.

Laura and her team began by conducting a thorough analysis of the production process at Sweet Dreams. They observed every step, from order receipt to product delivery. They meticulously recorded production times, identified areas of inefficiency, and sought opportunities for improvement.

What they found was enlightening. There were bottlenecks in ingredient preparation, lack of organization in order management, and extended delivery times. Additionally, the communication process between the kitchen team and the sales staff was inefficient, leading to errors and production delays.

With data and observations in hand, Laura and her team began implementing significant changes in the bakery. One of the initial changes was the reorganization of the ingredient preparation area. A more efficient storage system was established, and inventory management was improved to ensure that ingredients were always available in the right quantities. They also implemented a more efficient order management system. A clear workflow was established to process orders, from receipt to preparation and delivery. Specific responsibilities were assigned to team members to ensure efficient order handling.

Furthermore, communication between the kitchen team and the sales staff was improved. Regular follow-up meetings were established, and a visual management system was implemented to keep everyone informed about order status.

**Amazing Results:** On-Time Deliveries and Increased Customer Satisfaction

The results of implementing Lean at Sweet Dreams were astounding. The bakery managed to significantly reduce delivery times, consistently meeting delivery deadlines. This improved customer satisfaction and generated positive feedback that attracted new customers.

Efficiency in production also resulted in reduced operating costs. Sweet Dreams was able to uphold its commitment to high-quality ingredients and staff training while maintaining healthy profit margins.

**Lessons Learned:** The Importance of Efficiency and Communication

Laura and Sweet Dreams' story is a testament to the power of efficiency and communication in a business. Laura learned that to grow and remain competitive, it was crucial to optimize processes and improve communication within her team.

The bakery went from facing delivery delays and quality issues to being known for its punctuality and consistent quality. Customers trusted that their orders would be handled efficiently and that they would receive high-quality products.

**Conclusion:** A Sweet Future for Sweet Dreams

Today, Sweet Dreams remains a renowned bakery in the city. Laura and her team have upheld their commitment to continuous improvement and production efficiency. Laura's story and her boutique bakery serve as a reminder that, regardless of the industry, the application of Lean principles can lead to remarkable results. With dedication, vision, and a steadfast focus on excellence, any business can transform and thrive in an increasingly competitive market.

## Story 5: "Optimizing Customer Experience in a Call Center"

In the heart of the bustling city, in a modern office building, was the call center "ComuLine." This company provided customer service solutions for a variety of businesses in sectors ranging from telecommunications to e-commerce. While ComuLine had a strong customer base, it was struggling with a critical challenge: customer dissatisfaction due to extended wait times and inconsistent service quality.

Carlos Mendoza, the CEO of ComuLine, had been at the helm of the company for several years. He had witnessed its growth and expansion but had also observed how customer satisfaction issues were becoming increasingly prominent. Competition in the call center industry was fierce, and Carlos knew that to stay competitive, they needed to take bold steps.

**The Challenge:** Customer Satisfaction and Operational Efficiency

The primary challenge Carlos faced was customer dissatisfaction due to extended wait times and inconsistent service quality. Customers complained about long wait times to speak to a customer service representative and ineffective responses to their issues and inquiries.

These problems not only affected customer satisfaction but also led to high levels of employee turnover among the customer service agents. Agents felt overwhelmed by the volume of incoming calls and the pressure to resolve issues quickly.

**The Lean Solution:** Focus on Efficiency and Customer Experience

Carlos had heard about Lean methodology through a colleague who had applied these principles in his manufacturing company. Although Lean initially seemed more oriented toward manufacturing goods, Carlos was convinced that Lean principles could also be applied to the call center industry to address operational challenges and improve the customer experience.

Carlos decided it was time to act and apply Lean principles at ComuLine. The first step he took was to gather his management team and share his vision of improving efficiency and service quality. Everyone was willing to try something new to help the company overcome its challenges.

Carlos and his team began by conducting an in-depth analysis of the processes at ComuLine. They observed every step, from call reception to issue resolution. They meticulously recorded wait times, identified areas of inefficiency, and sought opportunities for improvement.

What they found was enlightening. There were multiple bottlenecks in the customer service process. Incoming calls were distributed unevenly among the agents, leading to extended wait times for some customers and an overload for others. Additionally, issue resolution procedures were not standardized, resulting in inconsistent responses.

With data and observations in hand, Carlos and his team began implementing significant changes at ComuLine. One of the initial changes was the reorganization of call distribution. They implemented an intelligent routing system that distributed calls more equitably among the agents, reducing wait times and uneven workloads.

Furthermore, they established standardized issue resolution procedures and provided additional training to agents to enhance their customer service skills. Clear guidelines were created to address a variety of issues and situations, ensuring consistent and effective responses.

**Amazing Results:** Increased Customer Satisfaction and Employee Retention

The results of implementing Lean at ComuLine were astounding. Customer satisfaction increased significantly due to reduced wait times and more effective responses from the agents. Customers began praising the improved service quality. Additionally, employee turnover among customer service agents decreased significantly. Agents felt less overwhelmed and more empowered to handle incoming calls, improving their job satisfaction and commitment to the company.

**Lessons Learned:** The Importance of Efficiency and Standardization

Carlos and ComuLine's story is a testament to the power of efficiency and standardization in a service-oriented business. Carlos learned that to enhance customer satisfaction and retain his staff, it was crucial to optimize processes and standardize procedures.

The company went from facing customer complaints and high employee turnover to being known for its excellent customer service and commitment to its team. ComuLine demonstrated that applying Lean principles can have a significant impact in the call center industry.

**Conclusion:** A Bright Future for ComuLine

Today, ComuLine is a leader in the call center industry, known for its efficiency and quality customer service. Carlos and his team have maintained their commitment to continuous improvement and customer satisfaction. Carlos's story and his company serve as a reminder that, regardless of the industry, applying Lean principles can lead to remarkable results. With dedication, vision, and a constant focus on excellence, any business can transform and thrive in an increasingly competitive market.

**Story 6: "Operational Efficiency in a Technical Support Call Center"**

In a modern business district of the city, you could find the call center "TechHelp." This company provided technical support services for a variety of technology companies worldwide. Despite having a loyal customer base, TechHelp faced critical challenges related to operational efficiency and customer satisfaction.

Ana López, the operations manager of TechHelp, had been at the helm of the call center for several years. She had risen from a technical support agent position and was familiar with the challenges and complexities of the industry. During her tenure as manager, she had seen the company grow and expand, but she had also noticed that issues related to efficiency and customer satisfaction were becoming increasingly prominent.

**The Challenge:** Operational Efficiency and Customer Satisfaction

The primary challenge Ana faced was operational efficiency and customer satisfaction. TechHelp had a highly skilled team of agents but wait times for customers in need of assistance were often long. This generated frustration among customers and negatively affected their experience with the company.

Furthermore, TechHelp's internal processes were not optimized. The assignment of cases to agents was inefficient, resulting in uneven response times and an overload for some agents and departments while others had a lighter workload.

**The Lean Solution:** Focus on Efficiency and Customer Experience

Ana had heard about Lean methodology through a fellow manager who had applied these principles in his financial services company. Although Lean initially seemed more oriented toward manufacturing goods, Ana was convinced that Lean principles could also be applied to the call center industry to address operational challenges and improve the customer experience.

Ana decided it was time to take action and apply Lean principles at TechHelp. The first step she took was to gather her management team and share her vision of improving efficiency and quality in the call center. Everyone was willing to try something new to help the company overcome its challenges. Ana and her team began by conducting an in-depth analysis of the processes at TechHelp. They observed every step, from call reception to technical issue resolution. They meticulously recorded wait times, identified areas of inefficiency, and sought opportunities for improvement.

They found that there were bottlenecks in the case routing process, a lack of standardization in issue resolution procedures, and ineffective communication between departments.

With the information at their disposal, Ana and her team began implementing significant changes at TechHelp. One of the initial changes was the reorganization of the case routing process. They implemented an intelligent routing system that assigned cases more equitably among the agents, reducing wait times and workload imbalances.

Furthermore, they standardized issue resolution procedures and provided additional training to agents to enhance their technical skills. A tracking and monitoring system was implemented to ensure cases were resolved in a timely and effective manner.

**Amazing Results:** Increased Efficiency and Customer Satisfaction

The results of implementing Lean at TechHelp were astounding. Operational efficiency significantly improved, resulting in shorter wait times for customers and a more equitable distribution of workload among agents. The company was able to consistently meet response deadlines, which improved customer satisfaction.

Additionally, the quality of technical support improved dramatically. Standardized procedures and additional training ensured precise and effective technical responses, reducing issue resolution times and increasing customer satisfaction.

**Lessons Learned:** The Importance of Efficiency and Communication

Ana and TechHelp's story is a testament to the power of efficiency and communication in a technical support call center. Ana learned that to improve customer satisfaction and retain her staff, it was crucial to optimize processes and enhance the quality of technical support.

The company went from facing long wait times and efficiency issues to being known for its efficient and effective technical support. TechHelp demonstrated that applying Lean principles can have a significant impact in the technical support call center industry.

**Conclusion:** A Promising Future for TechHelp

Today, TechHelp is a leader in the technical support industry, known for its exceptional efficiency and quality. Ana and her team have maintained their commitment to continuous improvement and customer satisfaction. Ana's story and her technical support call center serve as a reminder that, regardless of the industry, applying Lean principles can lead to remarkable results. With dedication, vision, and a constant focus on excellence, any business can transform and thrive in an increasingly competitive market.

**Story 7: "Reducing Delivery Times in an Automotive Parts Manufacturing Company"**

In a corner of an industrial city, you could find the automotive parts manufacturing company "AutoTech Solutions." This company specialized in producing a wide variety of automotive components, from engines to braking systems, serving car manufacturers nationwide. Despite having a solid reputation in the industry, AutoTech Solutions faced a critical challenge related to delivery times.

Maria Gonzalez, the Director of Operations at AutoTech Solutions, had a long history with the company. She started as a production engineer and worked her way up to her current position. During her time at the company, she had witnessed its growth and expansion, but she had also noticed that issues related to delivery times were becoming increasingly prominent.

**The Challenge:** Extended Delivery Times

The primary challenge Maria faced was extended delivery times. Despite producing high-quality automotive components, AutoTech Solutions often couldn't meet the agreed-upon delivery deadlines with car manufacturers. This led to customer dissatisfaction and put key business relationships at risk.

The lack of coordination between the production and logistics departments was a key issue. Orders were delayed due to a lack of visibility into internal processes and poor communication between teams.

**The Lean Solution:** Focus on Efficiency and Coordination

Maria had heard about Lean methodology through a colleague who had applied these principles in their manufacturing company. While Lean initially seemed more oriented toward goods production, Maria was convinced that Lean principles could also be applied to automotive parts manufacturing to address operational challenges and improve delivery times.

Maria decided it was time to take action and apply Lean principles at AutoTech Solutions. The first step she took was to gather her management team and share her vision of improving efficiency and coordination within the company. Everyone was willing to try something new to help the company overcome its challenges. Maria and her team began by conducting a thorough analysis of processes at AutoTech Solutions. They observed every step, from order reception to the delivery of finished products. They meticulously recorded production times, identified areas of inefficiency, and looked for opportunities to improve. They found bottlenecks in the production process, wasted time in material handling, and a lack of standardization in procedures. Additionally, communication between the production and logistics departments was deficient, leading to delivery delays.

Armed with data, Maria and her team began implementing significant changes at AutoTech Solutions. One of the initial changes was the reorganization of the production flow. They established more efficient production routes and implemented a visual management system so that workers could identify issues more quickly and effectively.

They improved coordination between the production and logistics departments by implementing regular meetings to review order statuses and address any issues promptly. A real-time tracking system was established for orders, allowing for greater visibility into internal processes and more agile decision-making.

**Amazing Results:** Reduced Delivery Times and Satisfied Customers

The results of implementing Lean at AutoTech Solutions were remarkable. Delivery times were significantly reduced, allowing the company to consistently meet delivery deadlines agreed upon with car manufacturers. This led to more satisfied customers and the strengthening of key business relationships.

Furthermore, production efficiency also improved, resulting in lower production costs and increased profitability for the company. AutoTech Solutions demonstrated that the application of Lean principles can have a significant impact on the automotive parts manufacturing industry.

**Lessons Learned:** The Importance of Efficiency and Coordination

Maria and AutoTech Solutions' story is a testament to the power of efficiency and coordination in an automotive parts manufacturing company. Maria learned that to improve delivery times and customer satisfaction, it was crucial to optimize processes and enhance coordination between departments.

The company went from facing extended delivery times to being known for its on-time delivery and customer satisfaction. AutoTech Solutions showed that the application of Lean principles can lead to a positive transformation in the automotive parts manufacturing industry.

**Conclusion:** A Promising Future for AutoTech Solutions

Today, AutoTech Solutions is a leader in the automotive parts manufacturing industry, known for its reduced delivery times and customer satisfaction. Maria and her team have maintained their commitment to continuous improvement and operational efficiency. Maria's story and her company serve as a reminder that, regardless of the industry, the application of Lean principles can lead to remarkable results. With dedication, vision, and a constant focus on excellence, any business can transform and thrive in an increasingly competitive market.

## Story 8: "Improving Quality in Automotive Parts Assembly Plant"

In a sprawling industrial area, you could find the automotive parts assembly plant "AutoParts Excel." This company specialized in manufacturing and assembling a variety of automotive components, from transmission systems to braking systems, which were supplied to renowned car manufacturers across the country. Despite having a solid reputation in the industry, AutoParts Excel faced a critical challenge related to the quality of its products.

Luis Morales, the Production Manager at AutoParts Excel, had been with the company for many years. He started as a line operator and worked his way up to his current position. During his time at the company, he had witnessed its growth and expansion, but he had also noticed that issues related to product quality were becoming increasingly prominent.

**The Challenge:** Quality Issues in Products

The primary challenge that Luis faced was the quality of the products manufactured at AutoParts Excel. Although the company produced highly complex automotive components, the rate of defective products had increased in recent years. This led to customer dissatisfaction and jeopardized key business relationships.

The lack of an effective quality management system and a lack of standardization in production procedures were key issues. Additionally, communication between the production and quality departments was poor, leading to late defect detection.

**The Lean Solution:** Focus on Quality and Standardization

Luis had heard about Lean methodology through a colleague who had applied these principles in their manufacturing company. Although Lean initially seemed more oriented toward goods production, Luis was convinced that Lean principles could also be applied to automotive parts assembly to address quality challenges and improve efficiency.

Luis decided it was time to take action and apply Lean principles at AutoParts Excel. The first step he took was to gather his management team and share his vision of improving product quality and efficiency in the plant. Everyone was willing to try something new to help the company overcome its challenges. Luis and his team began by conducting a thorough analysis of processes at AutoParts Excel. They observed every step, from the receipt of raw materials to the assembly of finished products. They meticulously recorded the rates of defective products, identified areas of inefficiency, and looked for opportunities to improve. There was a lack of standardization in production procedures, wasted time in material handling, and a lack of proper training for operators. Furthermore, communication between the production and quality departments was deficient, leading to late defect detection.

Armed with their observations, Luis and his team began implementing significant changes at AutoParts Excel. One of the initial changes was the standardization of production procedures. Clear work instructions were established, and additional training was provided to operators to ensure that everyone followed the same processes.

They improved coordination between the production and quality departments by implementing regular meetings to review product quality and address any issues promptly. A feedback system was established so that operators could report defects immediately, and corrective actions were taken promptly.

**Amazing Results:** Improved Quality and Satisfied Customers

The results of implementing Lean at AutoParts Excel were remarkable. The quality of the products improved significantly, resulting in a drastic reduction in defective products. Customers noticed the improvement in quality, and confidence in the company strengthened.

Furthermore, production efficiency also improved, resulting in lower production costs and increased profitability for the company. AutoParts Excel demonstrated that the application of Lean principles can have a significant impact on the automotive parts assembly industry.

**Lessons Learned:** The Importance of Quality and Standardization

Luis and AutoParts Excel's story is a testament to the power of quality and standardization in an automotive parts assembly plant. Luis learned that to improve product quality and plant efficiency, it was crucial to standardize processes and improve communication between departments.

The company went from facing quality and efficiency issues to being known for its high-quality products and production efficiency. AutoParts Excel showed that the application of Lean principles can lead to a positive transformation in the automotive parts assembly industry.

**Conclusion:** A Promising Future for AutoParts Excel

Today, AutoParts Excel is a leader in the automotive parts assembly industry, known for its product quality and production efficiency. Luis and his team have maintained their commitment to continuous improvement and customer satisfaction. Luis's story and his company serve as a reminder that, regardless of the industry, the application of Lean principles can lead to remarkable results. With dedication, vision, and a constant focus on excellence, any business can transform and thrive in an increasingly competitive market.

**Story 9: "Optimization in Automotive Parts Production: A Path to Efficiency"**

In the heart of a rapidly growing industrial zone stood the factory "AutoParts Pro." This company specialized in the production of a wide range of automotive parts, from engine components to braking systems, supplying renowned car manufacturers across the country. Although AutoParts Pro had a solid reputation in the industry, it faced significant challenges related to production efficiency and inventory management.

Carlos Rodríguez, the Director of Operations at AutoParts Pro, had been involved with the company for over two decades. He had climbed the ranks from a position on the production line and was well-versed in the intricacies of automotive parts manufacturing. During his time as Director of Operations, he had observed the company's growth and expansion, but he had also noticed that issues related to efficiency and inventory management were becoming increasingly pressing.

**The Challenge:** Production Efficiency and Inventory Management

The primary challenge Carlos faced was production efficiency and inventory management. AutoParts Pro prided itself on producing high-quality automotive parts, but production costs were constantly rising due to disorganized processes and extended production times. Additionally, the company struggled to maintain optimal inventory levels and adequately manage the fluctuating demand for its products.

The lack of coordination between the production and sales departments was also a problem. This resulted in a lack of real-time visibility into demand, leading to production delays and sometimes a shortage of products to fulfill customer orders.

**The Lean Solution:** Focus on Efficiency and Inventory Management

Carlos had heard about the Lean methodology through a colleague who had applied these principles in their manufacturing company. Although Lean initially seemed more geared toward manufacturing goods, Carlos was convinced that Lean principles could also be applied to automotive parts manufacturing to address their operational challenges and improve inventory management.

Carlos decided it was time to act and apply Lean principles at AutoParts Pro. The first step he took was to gather his management team and share his vision of improving efficiency and quality in the factory. Everyone was willing to try something new to help the company overcome its challenges.

Carlos and his team began by conducting a comprehensive analysis of processes at AutoParts Pro. They observed every step, from the receipt of raw materials to the delivery of finished products. They meticulously recorded production times, identified areas of inefficiency, and looked for opportunities to improve. There were bottlenecks in the production process, wasted time in material handling, and a lack of standardization in procedures. Additionally, communication between the production and sales departments was deficient, leading to production delays and a lack of visibility into demand.

Armed with their findings, Carlos and his team began implementing significant changes at AutoParts Pro. One of the initial changes was the reorganization of the production flow. More efficient production routes were established, and a visual management system was implemented to help workers identify issues more quickly and effectively.

Inventory management procedures were improved, and a more accurate demand forecasting system was implemented. This allowed the company to maintain optimal inventory levels and better anticipate fluctuations in demand. Furthermore, communication between the production and sales departments was improved, enabling more accurate planning and agile production.

**Amazing Results:** Greater Efficiency and Optimized Inventory Management

The results of implementing Lean at AutoParts Pro were astounding. Production efficiency improved significantly, resulting in shorter production times and lower costs. The company was able to consistently meet delivery deadlines, strengthening its relationships with car manufacturers.

Furthermore, inventory management was optimized, allowing AutoParts Pro to maintain optimal inventory levels and reduce costs associated with product storage. The company efficiently met demand without depleting inventory or incurring overstock.

**Lessons Learned:** The Importance of Efficiency and Collaboration

Carlos and AutoParts Pro's story is a testament to the power of efficiency and collaboration in an automotive parts factory. Carlos learned that to improve the company's profitability and inventory management, it was essential to streamline processes and enhance communication between departments.

The company went from facing efficiency and inventory management issues to being known for its efficient production and optimized inventory management. AutoParts Pro demonstrated that the application of Lean principles can have a significant impact on the automotive parts manufacturing industry.

**Conclusion:** A Promising Future for AutoParts Pro

Today, AutoParts Pro is a leader in the automotive parts manufacturing industry, known for its efficiency and exceptional inventory management. Carlos and his team have remained committed to continuous improvement and customer satisfaction. Carlos's story and his automotive parts factory serve as a reminder that, regardless of the industry, the application of Lean principles can lead to remarkable results. With dedication, vision, and a constant focus on excellence, any business can transform and thrive in an increasingly competitive market.

## Story 10: Transformation at an Auto Parts Factory, Improving Efficiency and Quality

On the outskirts of an industrial city stood the "AutoTech" auto parts factory. This company had been in operation for decades, providing a wide range of automotive parts to car manufacturers across the country. However, amid growing competition and economic challenges, AutoTech found itself at a crossroads.

Javier Sánchez, the Director of Operations at AutoTech, had a long history with the company. He had started his career in the factory as a production engineer and had risen to his current position. Over the years, he had seen the company grow and expand, but he had also witnessed issues related to efficiency and quality becoming increasingly prominent.

**The Challenge:** Production Efficiency and Quality Control

The main challenge Javier faced was production efficiency and quality control. AutoTech had a strong reputation for the quality of its products, but production costs were constantly rising due to disorganized processes and extended production times. Additionally, the company struggled to meet order delivery deadlines, which was affecting its relationship with car manufacturers.

Quality control had also become an issue. As production increased, it became increasingly difficult to maintain high-quality standards consistently. Defects and product returns were on the rise, resulting in additional costs and eroding AutoTech's reputation.

**The Lean Solution:** Focus on Efficiency and Quality

Javier had heard about the Lean methodology through a friend who had applied these principles in his manufacturing company. Although Lean initially seemed more geared toward manufacturing goods, Javier was convinced that Lean principles could also be applied to automotive parts manufacturing to address their operational challenges and improve quality.

Javier decided it was time to take action and apply Lean principles at AutoTech. The first step he took was to gather his management team and share his vision of improving efficiency and quality in the factory. Everyone was willing to try something new to help the company overcome its challenges.

Javier and his team began by conducting an in-depth analysis of processes at AutoTech. They observed every step, from the receipt of raw materials to the packaging of finished products. They meticulously recorded production times, identified areas of inefficiency, and looked for opportunities to improve.

What they found was enlightening. There were bottlenecks in the production process, wasted time in material handling, and a lack of standardization in procedures. Additionally, communication between the production and quality control departments was deficient, leading to the late detection of defects. Armed with these insights, Javier and his team began implementing significant changes at AutoTech. One of the initial changes was the reorganization of the production flow. More efficient production routes were established, and a visual management system was implemented to help workers identify issues more quickly and effectively.

Quality control procedures were improved, and additional training was provided to staff to enhance their inspection skills. A feedback system was implemented so that workers could report quality issues immediately, and timely corrective actions were taken.

**Amazing Results:** Greater Efficiency and Product Quality

The results of implementing Lean at AutoTech were astounding. Production efficiency improved significantly, resulting in shorter production times and lower costs. The company was able to consistently meet delivery deadlines, strengthening its relationship with car manufacturers.

Furthermore, product quality improved significantly. The implementation of more effective quality control procedures and early defect detection dramatically reduced defective products and returns.

**Lessons Learned:** The Importance of Efficiency and Quality

Javier and AutoTech's story are a testament to the power of efficiency and quality in a manufacturing business. Javier learned that to improve the company's profitability and reputation, it was essential to streamline processes and improve product quality.

The company went from facing efficiency and quality control issues to being known for its efficient production and high-quality products. AutoTech demonstrated that the application of Lean principles can have a significant impact on the automotive parts manufacturing industry.

**Conclusion:** A Promising Future for AutoTech

Today, AutoTech is a leader in the auto parts manufacturing industry, known for its exceptional efficiency and quality. Javier and his team have remained committed to continuous improvement and customer satisfaction. Javier's story and his auto parts factory serve as a reminder that, regardless of the industry, the application of Lean principles can lead to remarkable results. With dedication, vision, and a constant focus on excellence, any business can transform and thrive in an increasingly competitive market.

**Glossary**

Continuous Improvement: A fundamental principle of Lean that involves the ongoing search for ways to improve processes and eliminate waste.

Waste (Muda): Any activity or process that does not add value to the final product or service. Types of waste in Lean include overproduction, waiting times, unnecessary transportation, overprocessing, excess inventory, unnecessary motion, and defects.

Value: What the customer is willing to pay for a product or service. Creating value is the central goal in Lean.

Value Stream: The complete set of activities required to create a product or service from start to finish, including all necessary steps and processing time.

Kaizen: A Japanese concept that means "continuous improvement." It refers to the philosophy of seeking incremental and sustainable improvements in all aspects of an organization.

Just in Time (JIT): A Lean principle that focuses on producing and delivering products just when they are needed, eliminating the storage of unnecessary inventory.

Takt Time: The available time to complete one unit of production and meet customer demand. It is used to establish the production pace based on demand.

Kanban: A visual system used in Lean to control production and material flow. Kanbans are cards or signals that indicate when to produce or replenish products.

Poka-Yoke: A term referring to devices or methods that prevent errors in the production process. It is also called "error-proofing."

Work Cell: A production area where a group of workers performs all the operations necessary to produce a product or a family of products, eliminating unnecessary movements and waiting times.

Andon: A visual system that uses lights or indicators to signal problems or abnormalities in the production process.

Heijunka: Production leveling to maintain a constant flow of products over time, avoiding peaks and valleys in production.

Jidoka: A concept that refers to automation with a human touch. It involves giving machines the ability to automatically stop when a problem is detected, allowing workers to intervene and resolve the issue.

KPI (Key Performance Indicators): Metrics or measures used to assess the performance and efficiency of a process or organization.

Gemba: A Japanese term that refers to the actual workplace, where actions occur, and products are created. In Lean, the importance of being in the gemba is emphasized for understanding and improving processes.

## Epilogue

As we reach the end of this book, "What Is That Thing Called Lean? Stories of Continuous Improvement," we have journeyed through a world of transformation, innovation, and constant improvement. We have explored eight inspiring stories that span different industrial sectors, from poultry farming to auto parts manufacturing. In each of these stories, we have witnessed how the Lean philosophy has left an indelible mark on how we operate, improve, and thrive in our businesses and lives.

Each story presented unique challenges and seemingly insurmountable obstacles, but in each case, brave individuals and dedicated teams have shown that continuous improvement is possible in any setting. Through the application of Lean principles, they have achieved astonishing results: from reducing delivery times in an auto parts plant to operational efficiency in a technical support call center. But beyond the individual stories, these pages contain valuable lessons. We have learned that continuous improvement is not just a methodology, but a mindset rooted in the constant pursuit of excellence. We have discovered the importance of eliminating waste, standardizing processes, and giving workers the tools and autonomy to make a difference.

We have also explored fundamental Lean concepts, from value stream and takt time to kaizen and waste elimination. These concepts are not only applicable in the business realm but can also inspire us to seek continuous improvement in our personal and professional lives.

I want to thank each one of you, dear readers, for joining me on this journey. This book would not have been possible without your interest and desire to learn and grow. My family, my wife Yissel, and my son Santiago have been a constant source of inspiration, and I thank them for their unwavering support on this journey. As you conclude these pages, I invite you to carry with you the lessons of Lean and a passion for continuous improvement in everything you do. Remember that no matter how challenging the obstacles may be, there is always a way to move towards excellence. Lean teaches us that there is no end to improvement, and this book is a testament to that powerful principle.

May the stories shared here inspire and motivate you to embrace continuous improvement in all areas of your lives. The path to excellence begins with a single step, and every small improvement brings us a little closer to our goals.

With gratitude,

Javier Olivares.